I0467596

Do not worry about your difficulties in Mathematics. I can assure you mine are still greater.

Albert Einstein

(Letter to high school
student
Barbara Lee Wilson,
7 January 1943)

Preface

I am doing this compendium of articles, which I have been publishing for the last year, thinking in my students and my own children. They feel frustrate when they ask to their teachers or parents about real necessity of Mathematics, and the answer received is a group of abstract phrases that drive to more doubts instead clarify the necessity of Mathematics, even if you are thinking in to study Humanities, Laws, or other careers that look very far away from Math's field.

Mathematics helps to explain the world all around. Math is a powerful tool that let us discover the beautiful edges of the universe. Math is a path to grow as person and as a professional.

This group of articles is to encourage students to study math, through the explanation of real life and simple problems, mainly relate to economy and social issues.

I hope that through these pages you discover the importance of Mathematics, and feel more motivated about this wonderful science.

Prof. Felix Ramos, PhD in Engineering

Ex Assistant Professor of Engineering
University of Las Villas, Santa Clara, Cuba

Adjunct Faculty of Math and Statistics
MDC, Miami, Florida

Mathematics: Enemy or ally?

Mathematics' teachers hear frequently phrases such as: "... Teacher, I hate math ...." "... Math is not a very good friend ...." "... I always had problems with Math ..." or the most terrible of all: "... why do I need math...?"

All these phrases show a much deeper problem than a mere rejection of a group of students to a particular subject: **Our children are taught mathematics throughout their school life, but most of them can not perceive the relationship and the role of this science in daily activity beyond their trade or profession.** What is the cause? Is the inability of teachers who came before us? Is the treatment of the subjects in textbooks? Are deficiencies in the design of the curriculum? I do not pretend to answer these questions, because I have no evidence to enable me to reach a fair and objective response. I invite scholars of pedagogy to unravel the hidden reason for this rejection, towards this science, that has an essential role in the scientific development of our country and in our personal growth.

In my opinion, the role of a professor of mathematics goes beyond transmitting knowledge that some students feel as indigestible and boring; the teacher is responsible for finding motivational tools to turn this science in an interesting matter, to discover the beauty hidden behind the numbers and symbols, is in charge of guiding the youth in the development of their logical and analytical thinking. We have to feel the need to guide our students such that one of these days, casually, they approach us and say, "... Teacher, mathematics is really important in my life ...."

Through this article I want to present some very simple examples, related to daily life, showing us the importance of mathematics.

Let's start with the problems related to the calculation of percentages:

Consider the following scenario: _This weekend we're going to go shopping. We need some shirts, pants and a pair of shoes. We have received at home some of these discount coupons that usually come through the mail. One of the coupons is a "20% off" on a purchase over $50.00 and the other is "$15.00 off" on a purchase of equal magnitude. Suppose that when you buy the things needed, the total amount to pay is $68.56. The question is what discount coupon let us save more money?_

The following table shows the analysis we must do:

Amount to pay	Discount's coupons used	Saved amount	Amount to pay (after apply discount)
$68.56	$15.00 off	$15.00	$68.56 – $15.00 = $53.56
$68.56	20% off	20% of $68.56 is calculated as: 0.2($68.56) = $13.71	$68.56 – $13.71 = $54.85

As shown in the table above, in this scenario using the coupon "$15 off" you paid less money (we can save up to $1.29 compared to the "20% off" coupon).

What happens if the amount of our purchase is $137.58 now?

Amount to pay	Discount's coupons used	Saved amount	Amount to pay (after apply discount)
$137.58	$15.00 off	$15.00	$137.58 - $15.00 = $122.58
$137.58	20% off	20% of $137.58 is calculated as: 0.2($137.58) = $27.52	$137.58 - $27.52 = $110.06

In this case, using the coupon for "20% off", we will save a lot more money ($ 12.52 compared to the "$15.00 off" coupon). The explanation for this result is that the higher the cost of our purchase, the more money (discount) represents the "20% off" coupon, while the other is a fixed coupon discount of $15.00.

Consider a second example, also related to working with percentages: *As part of an investment, we bought shares in a company "X" listed on the stock exchange. Suppose you buy shares with a value of $*

120.00 per share. For certain reasons the market value of our shares fell by 20%, but we do not sell, we keep waiting for an upturn and this happens, so that after a time the value rises by 22%. The question is: Have our shares more, equal or lesser value that at the time of purchase?

Most of people might think that after an increase of 22% (greater than 20% reduction) our shares have more value. Will be this true?

The table shown below shows the analysis to be performed:

Initial value of the share	Value of the share after a decrease of 20% of its value	Value of the share after an increase of 22% of its value
$120.00	The 20% de $120.00 is computed as $0.2(\$120.00) = \$24.00$ value after the decrease $\$120.00 - \$24.00 = \$96.00$	The 22% de $96.00 is computed as $0.22(\$96.00) = \$21.12$ value after the increase $\$96.00 + \$21.12 = \$117.12$

As you can see, although the percentage increase in value was greater than the loss, the final value of our shares is lower (less than $2.88 per share). This is because to calculate the new value of the shares after the increase in 22%, we use the value of the stock after the fall of 20%, which is $96.00 instead of $120.00 initial value.

We intend to continue publishing, periodically, many more examples of the impact of mathematics in everyday life (interest paid for borrowings, using statistics to predict the probability of occurrence of an event, maximize or minimize certain magnitudes, etc.) so that parents and teachers, working in a coordinated way, motivate our children and encourage them into taste a science that does not age and that will define in a near future, our position as a scientific and technological power.

I want to buy a house: Math goes in support of the American's dream.

Everyone in America wish, one of these days, to buy their his/her own house; decorate it according to his/her style and put aside the anxiety of paying an "infinite" rent to live in a place that will never be yours. This act has been for decades, an essential part of the "American Dream."

To make financial decisions about buying a home, it is essential to be aided by a specialist in the subject, to help us choose the best decision for us; but it is also important to understand some basic principles that make us a "smart shopper", that means, a buyer who can provide criteria and understand the basis of each of the recommendations of our advisor. For this, the math comes to our aid. Consider some important definitions, which should be familiar to try to buy a house:

Mortgage: A long-term loan (maybe up to 30 or 40 years), with the purpose of buying a house in which the property is collateral for the loan. If you fail to pay the loan, the lender takes possession of the property.

Down Payment: The portion of the sales price of the home, which the buyer must pay to the seller initially. This down payment is a percentage of the sale price.

Amount of Mortgage: The amount borrowed, and is calculated as the difference between the sales price and the down payment. It is used to calculate the monthly payment.

Monthly payment: The payment that must be made monthly, throughout the life of the loan.

Many times, when buy; we are more concerned about the amount of the monthly payment that the final amount you will end up paying for the purchase. The average citizen in these days of crisis, almost live "paycheck to paycheck", that means, has a very tight budget and cannot exceed its expenses.

The monthly payment depends on several factors which include: amount borrowed, interest rate (interest rate), the time of the loan (time of the mortgage, etc.). The loan may have a fixed or variable rate of interest. When the interest rate is fixed, you make the same monthly payments throughout the life of the loan. For variable interest rates, the amounts payable change over time depending on changes in interest rates.

Some buyers qualify for loans insured by the "Federal Housing Administration" (FHA) and through these programs the down payment is much lower than for regular loans.

Most lending institutions require the buyer to pay one or more points at the time the loan begins. For example, a point involves a payment of 1% of the loan. This payment is made only once.

The following table (fragment) shows the monthly payment for every $1,000.00 of loans (including only the principal and interest)

Rate	Number of years					
%	15	20	25	30	35	40
6.5	$8.71	$7.46	$6.75	$6.32	$6.04	$5.85
7.0	8.99	7.75	7.07	6.65	6.39	6.21
7.5	9.28	8.06	7.39	6.99	6.74	6.58
8.0	9.56	8.36	7.72	7.34	7.10	6.95
8.5	9.85	8.68	8.05	7.69	7.47	7.33

Consider the following scenario, which will help us understand the decision-making process for the purchase of the house: _You have the opportunity to get a loan to buy a house, with the following conditions: loan of $200,000.00; a down payment of 10% and paying two points at closing._

The lender has the following options:

Option 1: To fund the purchase of the house with fixed interest rate of 8.5%, payable in 30 years. According to the table above, it would mean a monthly payment of $7.69 per each $1,000.00.

14

Option 2: To finance the purchase of the house with a fixed interest rate of 7%, payable in 20 years. This represents $ 7.75 per each $1,000.00.

Which option accepts? To decide we "do" some numbers. In both cases:

The initial payment will be 10% of the loan: $200,000.00 × 0.1 = $20,000.00. The amount to be financed will be equal to $200,000.00 - $20,000.00 = $180,000.00.

Closing costs (for two points is 2%) is 0.02 × $3,600.00 = $180,000.00 (this amount is not deducted from pay).

The following table, summarize the calculations of the monthly payment, total of all payments and the total money we pay for interest.

As shown in the table bellow, the interest paid on the 30 year financing ends up being 176.84% of the loan of $180,000.00; while the 20-year financing, **interest** represents 86% of the loan. On the other hand, the monthly payment was increased only very slightly to shorten the time of the loan.

Option	Monthly Payment ($)	Total of all the payments ($)	Interest to pay due to the loan ($)
1	(7.69×180,000)/1000 = 1,384.20	(1,384.20×30×12) = 498,312.00	318,312.00
2	(7.75×180,000)/1000 = 1,395.00	(1,395.00×30×12) = 318,312.00	154,800.00

This example, which we have developed purely for educational purposes, shows that Option 2 is better, because an increase of just over $10.00 in the monthly payment will represent in the long-term $163,512.00 in savings that you would pay in respect of interest (option 2).

Always seek advice when in a embarking in a project as important as buying a home, but we did not exclusively guide by the recommendations made by our legal advisors; we must take a pen and a piece of paper and make our own numbers to make sure that our decision is correct. Have no fear. The Mathematics is always available to help us.

Hamilton Circuits and Nearest Neighbor Method: Mathematical tools to save fuel and time.

In this article we want to continue to provide examples of the use of mathematics to achieve personal and professional success. This time we explain a mathematical method that can be used to optimize our routes and thus save fuel and time.

Our cities abound in small businesses offering home services: plumbing, electrical, air conditioning, delivery, among others. Many of these small businesses do not go beyond having a car or truck which transported, along with their tools and move all throughout our county, and even surrounding counties.

For the type of work they do (often simple repairs or with medium complexity, fast deliveries) can visit several clients, separated by tens of miles in a day's work. These are companies that do not generate large revenues and any small saving money and/or time can mean the difference between success and failure.

The path optimization can avoid us go unnecessarily tens of miles a day and thus save hundreds of dollars in monthly fuel consumption and time spent.

Consider the following scenario: _You have planned for tomorrow cleanups five air conditioners in different parts of the city, located in places as varied as: Doral, Miami, Hialeah, Cutler Bay and Kendall. The question here is: in what order should I place my route so as to minimize the miles to go?_

The first is to place the addresses of each site in a tool available to all, like Google Map. We find the distance between each of these points (we always offer more than one option, but we will choose the lower path involves. Another factor to consider when choosing the path between two points is the schedule and the density of traffic of that route). When it is finished we will get a map as I show in a figure bellow and from that map will create a complete graph (second figure)

A graph is considered complete when each pair of vertices are joined by a straight line. The vertices coincide with the sites where we will make our work. Lengths assigned to each axis, is the distance between each location in one of the remaining workstations.

Call **Hamilton circuit** will draw a path, such that we get out of a point or vertex, go through all vertices once and return to the starting point.

The number of circuits that form Hamilton complete graph are (n - 1)!, Where n is the number of vertices. As we have places to visit, we have (5-1)! possible circuits, this involves 4! = 4x3x2x1 = 24. Means there are 24 different ways to make our route. If we included only one more place to visit, the possible routes would raise to 60 combinations for testing (they take too long). The method analyzes all paths called "Brute Force Method" and is too cumbersome even for a supercomputer when the vertices become 15, 16 or more. It is the only really accurate method to find the optimal route.

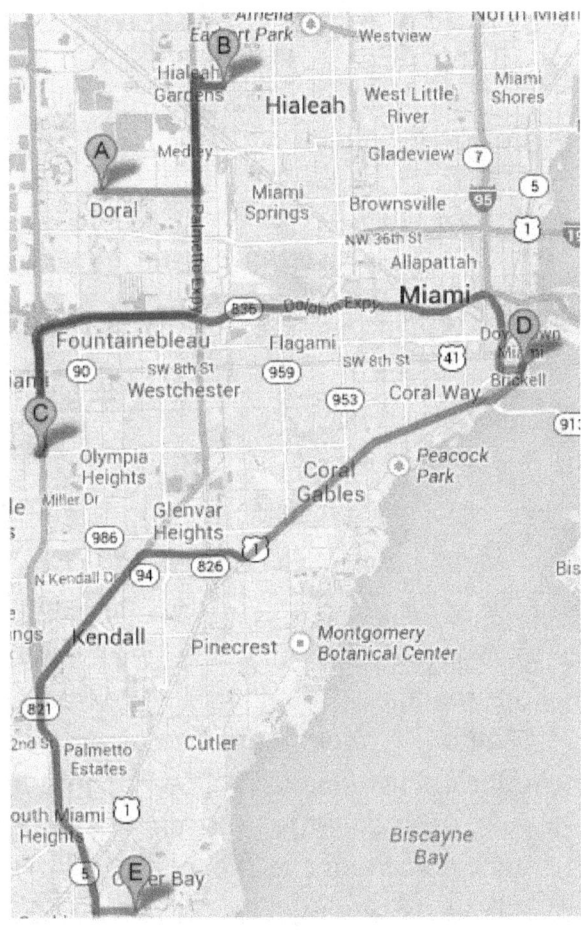

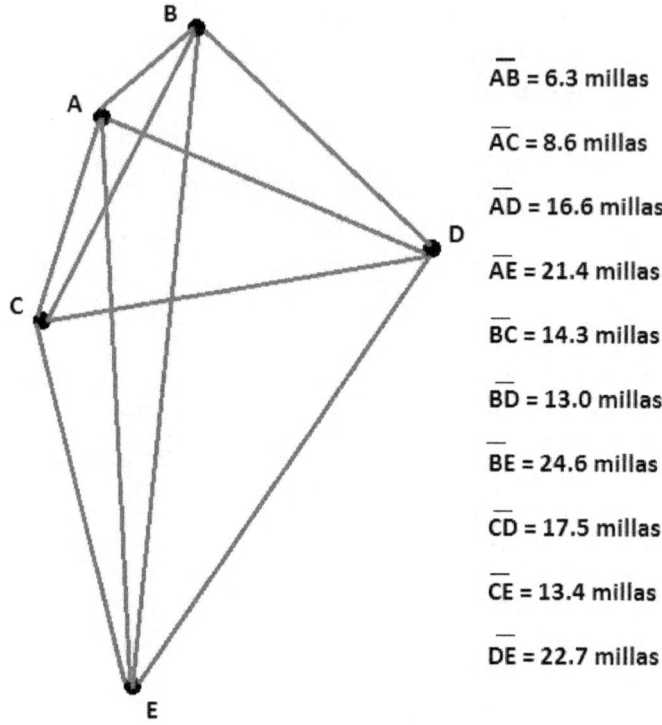

$\overline{AB}$ = 6.3 millas

$\overline{AC}$ = 8.6 millas

$\overline{AD}$ = 16.6 millas

$\overline{AE}$ = 21.4 millas

$\overline{BC}$ = 14.3 millas

$\overline{BD}$ = 13.0 millas

$\overline{BE}$ = 24.6 millas

$\overline{CD}$ = 17.5 millas

$\overline{CE}$ = 13.4 millas

$\overline{DE}$ = 22.7 millas

To avoid this situation we will explain an approximate method: "Nearest Neighbor Method." In our case we leave the vertex A (Doral). In this corner "A" segments AB, AC, AD, AE converge; the smaller length AB, so we'll go straight to the address in Hialeah (6.3 miles). Standing in "B", we find that the segments BC, BD, BE converge; the BD is shorter, so our next stop will be the city of Miami (13.0 miles). Now on "D" we find that we have two options: DC, DE; the shortest distance is by moving to Kendall, point C (17.5 miles). The last place to visit is Cutler Bay (following the EC path length 13.4 miles) and return to Doral (point A) through EA segment (21.4 miles). Our circuit is then as A, B, D, C, E, A (shown in Figure 2) and total travel miles to be 6.3 + 13.0 + 17.5 + 13.4 + 21.4 = 71.6 miles.

What happens if we choose randomly any other circuit? Let's see:
A, B, C, D, E, A = 82.2 miles (10.6 miles more to go).
A, C, B, D, E, A = 80.0 miles (8.4 miles more to go).

A, D, E, C, B, A = 73.3 miles (1.7 miles more to go).

However, the drawback of this method is that even going to offer a shorter journey than most of the possible paths, this will not necessarily be optimal. In this case a better way is to go by:

A, C, E, D, B, A = 64 miles (7.6 miles less to go, on the road offered by the Nearest Neighbor method). In this case it was easy to define this path less traveled, for the small number of sites to visit.

We can consider worth using this method despite its limitations because: it is very easy to implement (just Internet access and minimal skills with Google Maps) and we will find a path better, that most of the others possible.

System of linear equations: Tools for Business Success.

Many times we have wondered about, how can be used such systems of linear equations, that math teachers insist on forcing us to solve. Usually that question is school year, after school year, floating in the air and we graduated without understanding the utility of such "monsters" mathematicians.

Through this brief article, we try to discover together the importance of addressing and understanding the meaning of solving a system of equations, when applied to the relationship "costs - income" derived from the operation of a small business, producing definite article.

First let us see, briefly, some definitions of the field of economics:

Income (I) is calculated as the product of the number of items sold (X) and the unit selling price of the items (P). In a mathematical notation, the revenue function is written as:

$$I(X) = P \times X$$

Costs (C) is calculated as the sum of the fixed cost (referred to as CF) and the product produced unit cost (c) and the number of units produced (X); whereas each of the units we produce will be sold. The cost function is mathematically represented as:

$$C(X) = CF + c \times X$$

Fixed costs (FC) are those costs that the company will pay regardless of their level of operation. Included as fixed costs: paying the rent of facilities, payment of insurance, and in some cases, the cost of wages.

The **point of equilibrium** (breakeven point) is essential in this analysis. This point is graphically defined as the intersection point of the revenue function, I (X) and the cost function, C (X). From the economic point of view, the value of "x" coordinate of that point alerts us about how many units should be produce such that the money coming out of our company due to product costs, equals to the money coming in because income.

The following simple example illustrate this: _A small company produces shoes to a monthly fixed cost (CF) of $2,500.00. The cost of each produce each pair of shoes is $5.00 and the selling price of the shoes to retailers is $12.00. What is the minimum amount of shoes to be produced monthly to ensure that the company has profits (income exceeds expenditure)?_

The revenue function is:

$$I(X) = 12X$$

The cost function is determined as:

$$C(X) = 2500 + 5X$$

Using Microsoft Excel, we will calculate the values of the functions I (X) and C (X) for different numbers of produced pairs shoes, and graph both functions on a Cartesian coordinate system, as show the table and graph bellow:

Units produced and sold	Revenue	Costs
X	I(X)	C(X)
0	0	2500
100	1200	3000
200	2400	3500
300	3600	4000
400	4800	4500
500	6000	5000

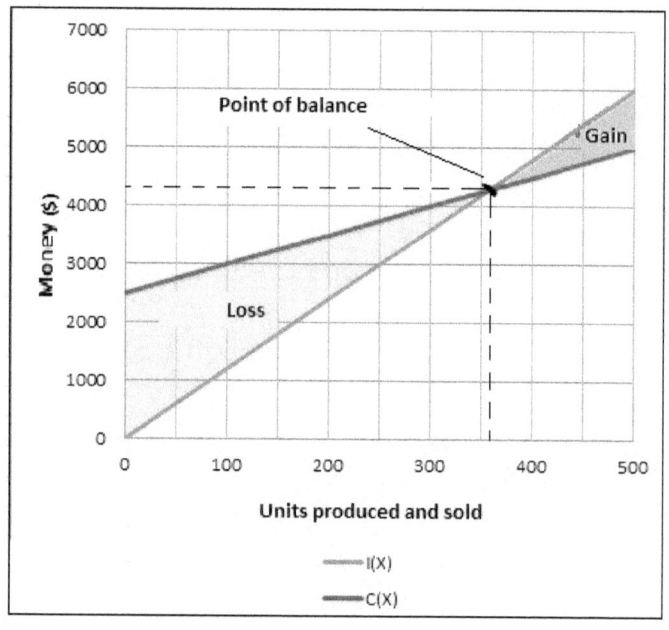

According to the graph, the company has earnings when produce and sell more than 360 pairs of shoes per month (x-coordinate of the point of equilibrium). As the production and sale exceed the above amount, the profit margin of the company will increase.

This simple example, developed for teaching purposes, shows that our math teachers are not torturing us, when ask us to solve systems of equations. They simply try to put powerful tools in our hands to reach the personal and professional success.

The systems of inequalities maximize the profits of your small business.

During the Second World War was developed a mathematical technique that allowed increasing the amount of inputs that were airlifted to supply the troops.

This technique called "Linear Programming" allows solving problems in which a certain amount must be maximized or minimized. In a very basic way, is the graphical or algebraic solution of a system of linear inequalities.

Before illustrate how this technique helps in the growth of a small business, we should mention a couple of important definitions.

Objective function is an algebraic expression in two or more variables describing the quantity to be maximized or minimized.

Constrains are conditions imposed by the real problem being modeled. Each constrain is written as a linear inequality, and grouped form a system of linear inequalities.

We are going to solve now a hypothetical problem, which may be faced by any small business owner.

A manufacturer of embroidered uniforms has a gain of $5.00 on each uniform (including pants and shirt / blouse), and $3.00 in each embroidered logo.

The manufacturer, due to the machines installed, number of employees and capital available, faces the following constraints:

1. The maximum amount of uniforms that can make monthly is 1000.

2. Maximum number of logos which could be embroidered is 800.

3. The cost of producing a uniform is $9.00 and make embroidery is $6.00. The monthly business expense cannot exceed $12,000.

To help maximize gains to our employer, we begin writing the objective function. In this function the variables used represent the following quantities: "X" is the number of uniforms produced in a month; "Y" is the number of embroidery made in one month; "Z" is the gain obtained by the company from its production (variable to maximize). The objective function becomes:

$$Z = 5X + 3Y$$

We are going to denote the constrains now, by using linear inequalities:

$$X \leq 1000$$

$$Y \leq 800$$

$$9X + 6Y \leq 12,000$$

Using Microsoft Excel, the inequalities are plotted and the system is solved, shading the area of the graph where all solution sets intersect, as shown in the figure bellow.

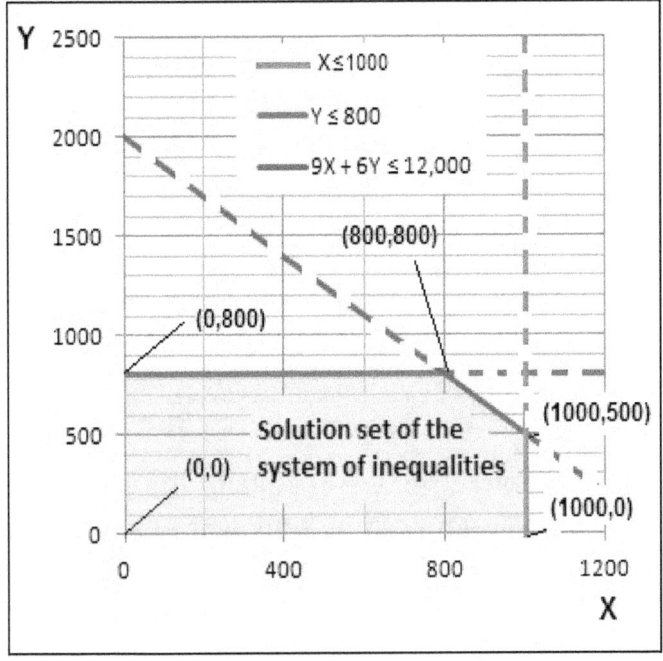

The shaded area is an irregular polygon of 5 sides and 5 vertices (intersection points between the sides). Each vertex corresponds to an X coordinate (number of uniforms produced) and a Y coordinate (number of embroideries). Substituting X and Y coordinates of each vertex in the objective function, the table below is obtained:

X Uniforms produced	Y Embroideries produced	Z Profit obtained $
0	0	0
1,000	0	5,000
1,000	500	6,500
800	800	6,400
0	800	2,400

According to the results of the table above, the maximum gain is achieved when the company produces 1,000 monthly, and 500 uniforms are embroidery. This maximum gain is $6,500 per month.

Now it is the employer, who must organize the personal and the business resources to ensure a maximum gain.

As you can see, we do not need great math skills to make our company more efficient.

Winning the Lottery: improbable Illusion?

Every day millions of people in our country play the lottery chasing an illusion: "Winning the jackpot."

This illusion is fueled every three, four or five weeks when we hear that someone, somewhere, won some millions of dollars, and went to swell the coveted group of millionaires.

Other moments that generate great excitement is when weeks and weeks pass without winners, and prizes begin to increase, reaching tens and sometimes, even, hundreds of millions. The population then turns frantically playing with a renewed hope of a windfall. I confess that I have spent too a few dollars on those occasions and I have eaten my nails as I wait awake the winning numbers. So far I have not gone beyond a free ticket.

How likely is to win the lottery jackpot? Let's see what the experts say. The following phrase was extracted from a famous book of probabilities and statistics, used in most colleges and universities in South Florida: **"The lottery is a tax on people who are bad in math."**

What is this approach? Is there really a contradiction between playing lottery game and the knowledge of mathematics? To find the answer let's dig a little bit in the calculation of probabilities.

What is probability? Let's find any online dictionary of the English language and look for the definition:

Probability:
1. F. Quality or credible and well-founded possibility that something might happen.
2. Mat. Calculation or quantitative determination of the possibility of an incident is verified.

In other words, the probability measures the likelihood that a particular event (for example, winning the lottery jackpot) occurs or not.

The probability (P) that an event occurs is a number between zero and one, such an event is impossible if its probability is zero, and an event will certainly occur if the probability is one.

Teachers of Statistics and Probability usually explain this by throwing a six faces dice numbered 1 to 6. What is the probability that you roll out a nine? Obviously is zero because there is no face numbered with nine. What is the probability that a number between one and six leave us? The answer is one, as these are the numbers on the faces of the dice.

An event will be considered improbable (probability of occurrence is very small) when the probability is a number equal to or less than 0.05.

After this brief reminder of the probability theory, we are almost ready to see how likely it is to win the lottery jackpot. We will base our calculation on two of the most popular games, "Fantasy 5" and "Lotto

Fantasy 5   Lotto

The "Fantasy 5" has numbers from 1 to 36 and you must select 5 of these numbers, without replacement (not allowed to repeat numbers in the same panel). As no matter the order in which the numbers go, the calculation of all possible different combinations is performed using the formula

$$_{36}C_5 = \frac{36!}{(36-5)!5!} = 376{,}992$$

Now if you play only one combination, your probability of winning is equal to dividing one by the number of different combinations, obtained from the above formula:

$$P_{(\text{ win })} = \frac{1}{376,992} \approx 0.000003$$

As you can appreciate the value of 0.000003 obtained is too much lower than 0.05 (value that delimits an unlikely event), so it is very likely that you are not the winner. In the "Lotto" is even more difficult to win. You must select 6 numbers from 53 possible, without replacement. Calculating the number of combinations

$$_{53}C_6 = \frac{53!}{(53-6)!6!} = 22,957,480$$

If you play one combination:

$$P_{(\text{ win })} = \frac{1}{22,957,480} \approx 0.00000004$$

This likelihood is ridiculously small. No matter if you play 10, 50 or 100 combinations, your chance will remain much lower than 0.05 and therefore the victory is very improbable.

The purpose of this article is not to say categorically: "Do not play lottery". However, we must play responsibly. The lottery is not the way to solve your materials and /or spiritual problems. The only way to succeed is through the study, the acquisition of knowledge and skills, and dedication to work.

I confess, I keep playing some numbers occasionally, but most of my time and my money will be invested in study and become more competent every day.

Are fair the democratic voting systems?

In 1951 the economist Kenneth Arrow (Nobel laureate in Economics in 1972) proved the validity of the now known as: **"Impossibility Theorem of Arrow"**. This theorem states that "it is mathematically impossible for any democratic voting system, satisfy the four criteria of fairness." In other words, does not exist, nor will there ever be an equitable democratic vote when the vote involve more than two candidates.

Which are the criteria of fairness that Arrow mentioned? The four criteria are:

Majority Criterion: If a candidate receives a majority of votes to place him first in an election, then he should be the winner of the election.

Criterion "Head to Head": If a candidate is favored when compared separately (meaning of "head to head") with each of the other candidates, then this should be the winner of the election.

Monotonicity Criterion: If a candidate wins an election and reelection, the only changes are changes that favor this candidate, this should be the winner of the election.

Irrelevant Alternatives Criterion: If a candidate wins an election and count the only changes are that one or more candidates are removed from the ballot, then the winning candidate must still be the winner of the election.

Here are some of the methods that are used to determine the winner of an election:

Plurality Method: The candidate with the most votes locates it first is the winner.

Borda Count Method: Candidates are organized in a more favored to least favored table, every last place gets one point each junior receives two points, sophomore receives three and so on. The candidate with the most points is the winner.

Plurality with Elimination Method: The candidate with the most first places wins. If no candidate receives a majority of first-place votes is eliminated from the candidate preferences table with the least amount of votes for first place and the rest of the candidates are moved up one place. If no candidate with a majority of first places, will be the winner, if the process is not repeated.

Pairwise Comparison Method: Using a table of preferences, each candidate is compared to the other candidates. In each comparison, the preferred candidate receives a point and if there is a tie, then receives midpoint. At the end of all comparisons the candidate receiving the most points will be the winner.

Take a simple example: _Let us assume that four candidates are running to be President for the Commission of a city "X". The names of the candidates are: Jorge (J), Pedro (P), Ramon (R) and Carlos (C). The city commission consists of 18 commissioners (including the four postulates and not participate in the vote)._ The preference table which contains the result of the ballot, of the fourteen commissioners voting is shown below:

According to the method of the plurality, the winner of the election would be Jorge receiving six votes in first place; Ramon is second with four first places; Carlos then three votes for president and finally Pedro with only one vote for the first position.

Preferences Table				
# of ballots with similar order	6	4	3	1
1st place	J	R	C	P
2nd place	R	P	R	J
3er place	P	J	J	R
4th place	C	C	P	C

We now apply the method of Borda count. We will assign four points for first-place votes, three second place votes, two votes in third place and finally a point for votes in fourth. Thus each candidate will receive:

According to the results obtained by this method is obviously Ramón who gets the highest score (forty-five points) and therefore won the vote.

Candidate	(Votes x Points for position)	Total points
Jorge (J):	(6x4) + (4x2) + (3x2) + (1x3) = = 24 + 8 + 6 + 3	41
Ramón (R):	(6x3) + (4x4) + (3x3) + (1x2) = = 8 + 16 + 9 + 2	45
Pedro (P):	(6x2) + (4x3) + (3x1) + (1x4) = = 12 + 12 + 3 + 4	31
Carlos (C):	(6x1) + (4x1) + (3x4) + (1x1) = = 6 + 4 + 12 + 1	23

By using two different systems for counting votes, we have obtained different winners. On the one hand, the plurality method (which assumes winner Jorge) violates the criterion of "head to head" for a candidate to compare against another Ramón received higher scores than Jorge. On the other hand, the Borda method (which assumes winner Ramon) violates the criterion of majority; Jorge has received the most votes for first place Ramon. If you test the other two methods mentioned at the beginning, perhaps we will find new winners.

Do these results indicate that democratic voting is an illusion, a fallacy?

Of course not. The issue is to establish the rules, clearly well before the election takes place. Moreover, when used either democratic voting system, it violates one or another of the principles of equity; but when officers are chosen by "finger", like the "totalitarian regimes", all together principles of equity are violated.

Sticking to the democratic vote, despite its inevitable shortcomings.

I hope that after reading and analyzing the above examples, you feel a little more motivated to study mathematics. This is an unparalleled tool in making personal and professional decisions.

The success is in your hands.

Prof. Felix Ramos

Author's Edition.
Miami, November 2015.
USA.

Author's contact information:
felixrm1971@gmail.com